# Planet EARTH

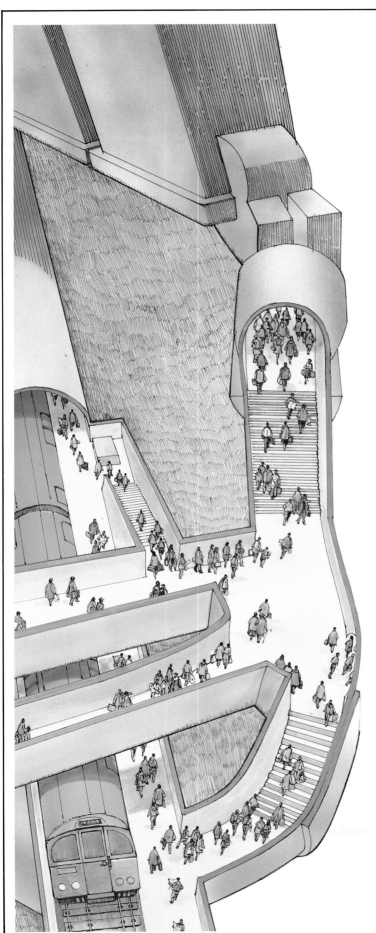

## Titles in this series:
Planet Earth
Rain Forest
Seas and Oceans
Stars and Planets

Editor: Deborah Biber
Illustrated by: Gerald Wood
Series Editor: David Salariya
Design Assistant: Steve Longdale
Project Manager and Electronic Production:
    Julie Klaus
Special thanks to Keith Lye for his assistance

**Library of Congress Cataloging-in-Publication Data**
Jessop, Joanne.
    Planet Earth / Joanne Jessop; illustrated by
Gerald Wood.
        p.    cm. — (New view)
    Includes index.
    ISBN 0-8114-9244-3
    1. Physical geography — Juvenile literature.
[1. Physical geography.]   I. Wood, Gerald, ill.  II. Title.
III. Series: New view (Austin, Tex.)
GB58.J47   1994
910'.02—dc20
                                                93-28339
                                                CIP
                                                AC


Printed and bound in Belgium.

1 2 3 4 5 6 7 8 9 0    99 98 97 96 95 94 93

# NEW View Planet EARTH

Written by
**JOANNE JESSOP**

Illustrated by
**GERALD WOOD**

Created and Designed by
**DAVID SALARIYA**

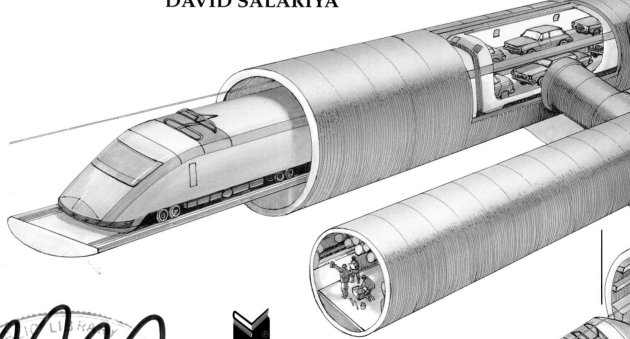

**RSVP**
**RAINTREE**
**STECK-VAUGHN**
**PUBLISHERS**
The Steck-Vaughn Company

*Austin, Texas*

# CONTENTS

## THE SHAPE OF THE EARTH

The Earth is not perfectly round. It is slightly flattened at the poles and has a bulge just south of the equator, making the planet Earth somewhat pear-shaped.

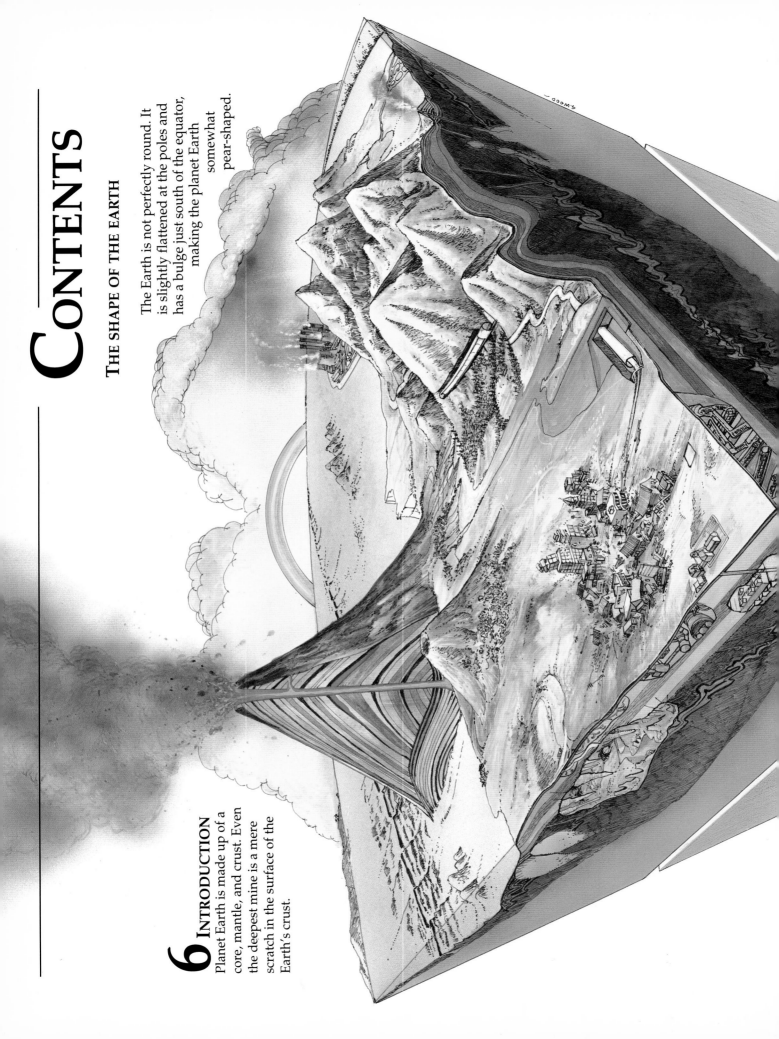

## 6 INTRODUCTION

Planet Earth is made up of a core, mantle, and crust. Even the deepest mine is a mere scratch in the surface of the Earth's crust.

# INTRODUCTION

Planet Earth consists of a central core surrounded by a mantle and a thin outer crust. What we see — the oceans, rivers, fields, and forests — is only a covering over the Earth's crust, which is made up of large rocky "plates" that include the crust and the top, solid part of the mantle. Most scientists believe that currents in the semi-molten layer of the mantle cause the plates to "float." As they drift apart and collide, mountain ranges are formed, volcanoes erupt, and earthquakes shake the ground. These powerful forces within the Earth are constantly changing its surface, but there are other forces at work as well. Soil is created as wind, rain, and ice erode the rocky crust. Growing plants and animals that burrow into the Earth's surface help in this process. The chemical action of rainwater falling on some types of rocks hollows out caves and creates underground rivers. People have made inroads into the Earth's crust with tunnels and mines.

This book takes a new view of planet Earth, starting with mountains, volcanoes, and earthquakes. It then proceeds down through the surface layer of the soil, with its array of animal inhabitants, to the further depths of the rocky crust riddled with caves, shafts, tunnels, and mines, and finally through the mantle to the inner core itself.

The cross section of planet Earth shown on the contents page appears in every chapter. The section that is highlighted indicates which particular layer of the Earth is now in view.

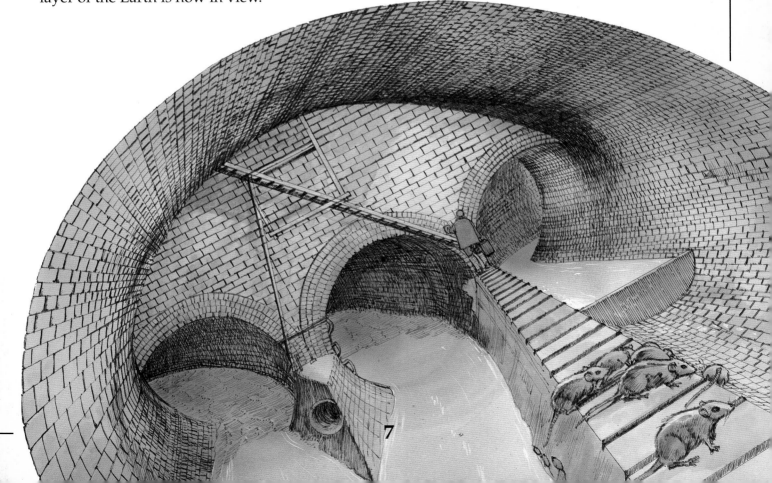

7

# MOUNTAINS

Mountains are created as the result of movement in the Earth's plates. Plates move very slowly, from less than one-quarter of an inch to 6.3 inches (.5cm to 16cm) a year. But over millions of years, the pressure of two plates pushing against each other can squeeze horizontal layers of rock between them into great folds.

A collision between an oceanic plate beneath the Pacific and the continental plate of South America caused volcanic activity that created the Andes mountain range about 65 million years ago. The Himalayan mountains began rising 80 million years ago when the plate supporting India pushed into Asia. The Alps, the world's youngest mountains, were formed when Africa was forced into Europe about 15 million years ago. All these mountain ranges are still growing as the plates beneath them continue to push against each other. The Scandinavian Mountains and the Appalachians, formed 250 to 400 million years ago, have stopped growing because the plates beneath them are no longer exerting any pressure.

①

②

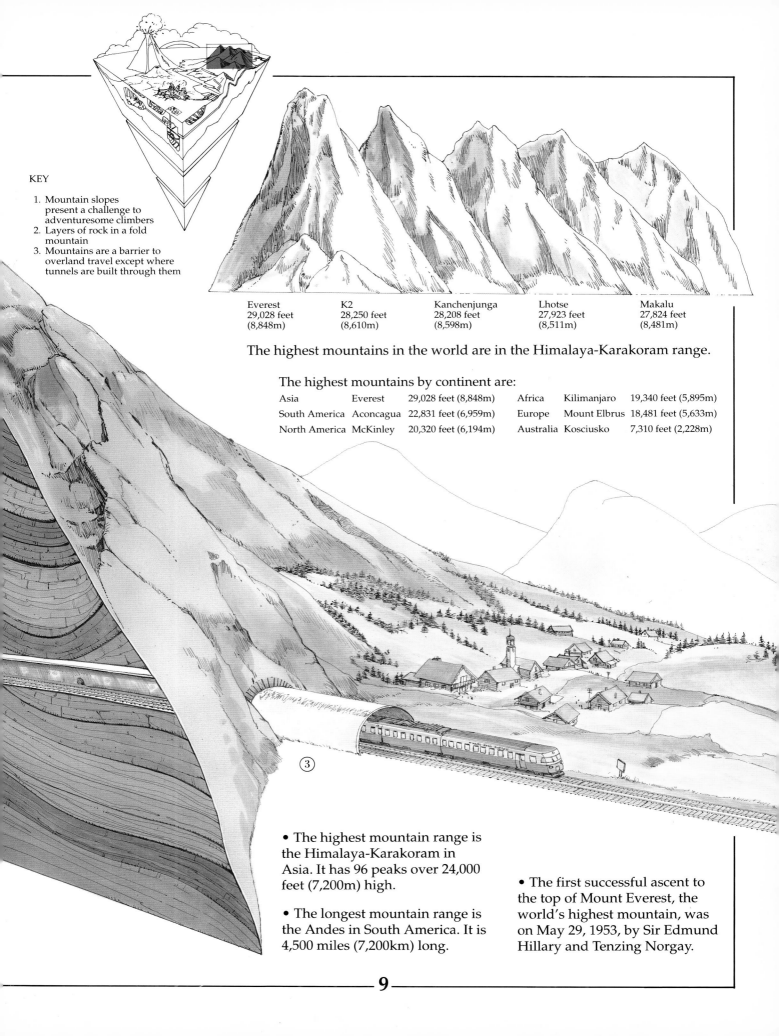

KEY

1. Mountain slopes present a challenge to adventuresome climbers
2. Layers of rock in a fold mountain
3. Mountains are a barrier to overland travel except where tunnels are built through them

| Everest | K2 | Kanchenjunga | Lhotse | Makalu |
|---|---|---|---|---|
| 29,028 feet | 28,250 feet | 28,208 feet | 27,923 feet | 27,824 feet |
| (8,848m) | (8,610m) | (8,598m) | (8,511m) | (8,481m) |

The highest mountains in the world are in the Himalaya-Karakoram range.

The highest mountains by continent are:

| Asia | Everest | 29,028 feet (8,848m) | Africa | Kilimanjaro | 19,340 feet (5,895m) |
|---|---|---|---|---|---|
| South America | Aconcagua | 22,831 feet (6,959m) | Europe | Mount Elbrus | 18,481 feet (5,633m) |
| North America | McKinley | 20,320 feet (6,194m) | Australia | Kosciusko | 7,310 feet (2,228m) |

- The highest mountain range is the Himalaya-Karakoram in Asia. It has 96 peaks over 24,000 feet (7,200m) high.

- The longest mountain range is the Andes in South America. It is 4,500 miles (7,200km) long.

- The first successful ascent to the top of Mount Everest, the world's highest mountain, was on May 29, 1953, by Sir Edmund Hillary and Tenzing Norgay.

9

# VOLCANOES

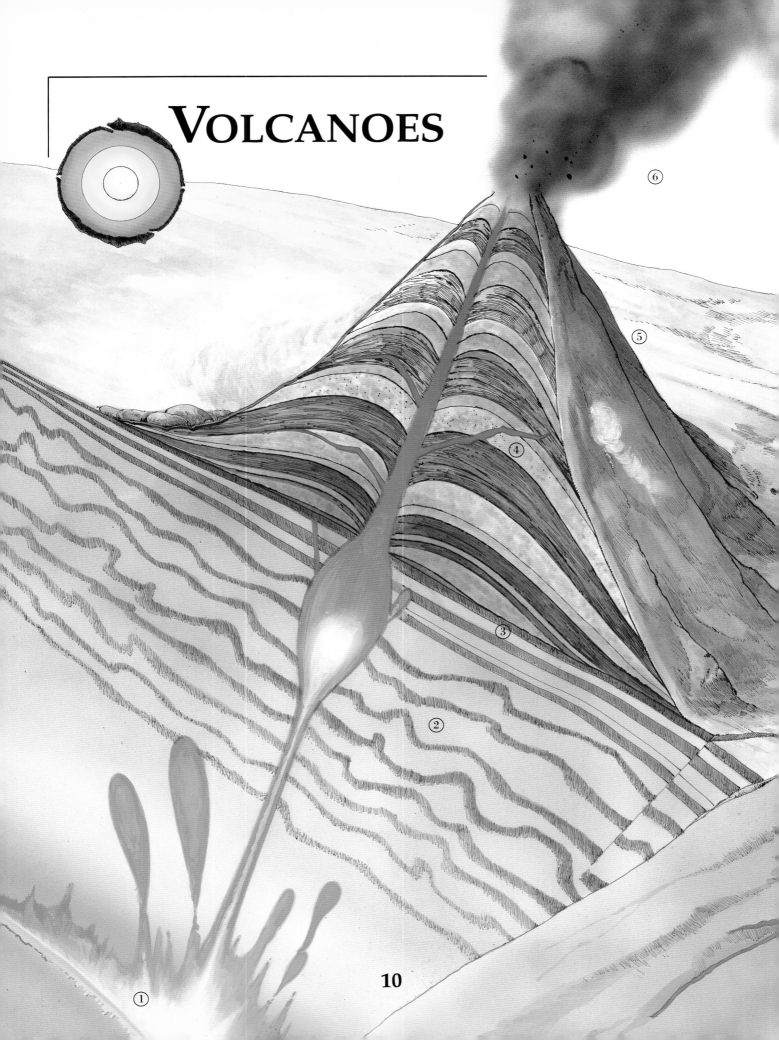

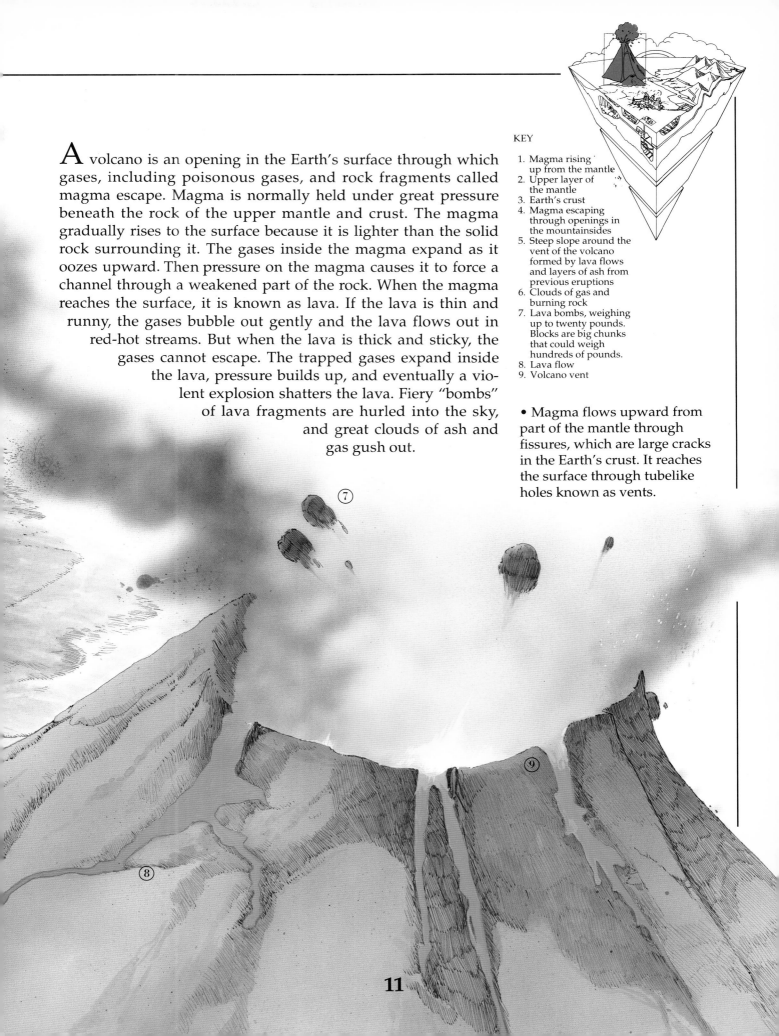

A volcano is an opening in the Earth's surface through which gases, including poisonous gases, and rock fragments called magma escape. Magma is normally held under great pressure beneath the rock of the upper mantle and crust. The magma gradually rises to the surface because it is lighter than the solid rock surrounding it. The gases inside the magma expand as it oozes upward. Then pressure on the magma causes it to force a channel through a weakened part of the rock. When the magma reaches the surface, it is known as lava. If the lava is thin and runny, the gases bubble out gently and the lava flows out in red-hot streams. But when the lava is thick and sticky, the gases cannot escape. The trapped gases expand inside the lava, pressure builds up, and eventually a violent explosion shatters the lava. Fiery "bombs" of lava fragments are hurled into the sky, and great clouds of ash and gas gush out.

KEY

1. Magma rising up from the mantle
2. Upper layer of the mantle
3. Earth's crust
4. Magma escaping through openings in the mountainsides
5. Steep slope around the vent of the volcano formed by lava flows and layers of ash from previous eruptions
6. Clouds of gas and burning rock
7. Lava bombs, weighing up to twenty pounds. Blocks are big chunks that could weigh hundreds of pounds.
8. Lava flow
9. Volcano vent

• Magma flows upward from part of the mantle through fissures, which are large cracks in the Earth's crust. It reaches the surface through tubelike holes known as vents.

11

• The first instrument for detecting earthquakes was invented by the Chinese in A.D. 132. When earthquake tremors shook the ground, a pendulum hanging inside the bronze pot caused one dragon's mouth to open, releasing a ball into the mouth of the frog below.

1) When areas of rock are pushed and pulled in opposite directions by shifting plates, a crack, or fault, develops.

2) That part of the crust may remain quiet for a long time, but eventually the same forces start bending the crust.

3) When the strain becomes too great, the rock snaps again, usually along the same fault.

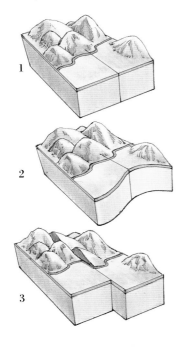

# EARTHQUAKES

As plates slide past and beneath one another, great stresses and strains build up. The rock of the crust can bend and adjust to a certain amount of strain. But when the pressure becomes too great, the rock snaps and shifts, causing the ground to shake. Vibrations, known as seismic waves, are sent out from the center of the earthquake and travel through the Earth. Some powerful seismic waves can be felt hundreds of miles away. There are thousands of earthquakes every year. Most are so small they pass unnoticed, but some quakes are so violent they tear open the ground and shift huge hunks of land. In heavily populated areas there may be many deaths and injuries as buildings crumble and roads and bridges collapse.

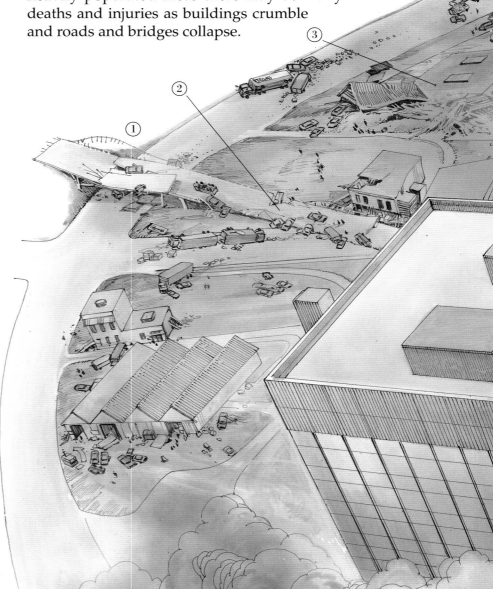

KEY

1. Road and bridges wrecked
2. Traffic accidents occur as cars fall into cracks in the road
3. Buildings collapse, trapping people inside
4. Clouds of choking dust
5. Building sinks as soil beneath liquefies
6. Fires break out; water and power supplies are cut off
7. Emergency services disrupted; the only access to the area is by helicopter

# SOIL

• Earthworms are like tiny plows. They can turn over about 1.5 tons of soil per acre of grassland in one year.

When the Earth first formed about 4.6 billion years ago, it was covered with blazing hot rock. Gradually it cooled down. Over millions of years, rain, wind, and ice eroded the surface of the rock. When organic matter in the form of decomposed plants and animal remains mixed with rock fragments, soil was formed. Soil is constantly being created and added to the Earth's surface. It can also be rapidly destroyed by careless farming and logging methods.

Soil contains many living things, which all help the soil in some way. Plant roots hold in moisture and prevent the soil from being blown or washed away. Ants, worms, centipedes, and other animals drag organic matter down into the soil, which, along with their body wastes, enriches the soil. As these animals feed on organic material, they start the decaying process. Bacteria in the soil continue the process by breaking down the organic matter into the mineral salts that plants need to grow.

KEY

1. Butterfly egg
2. Larva, or caterpillar
3. Chrysalis
4. Red admiral butterfly
5. Worm cast (soil brought to the surface by an earthworm)
6. Ant carrying leaf into its nest
7. Earthworm
8. Queen ant laying eggs
9. Fungus growing inside ant nest
10. Dead moth inside ant nest
11. Trapdoor spider
12. Dead beetle about to be carried underground by ants
13. Leather-jacket larvae
14. Caterpillars
15. Ladybugs
16. Wood louse
17. Caterpillar
18. Mouse skeleton
19. Tree roots
20. Earthworm
21. Wood-boring insect
22. Centipede
23. Red admiral butterfly
24. Snail

# ANIMALS UNDERGROUND

The quiet, peaceful scenery of the countryside conceals the bustle of activity going on beneath the ground. The soil is full of creatures burrowing in search of food. At the first sign of danger, many small mammals scurry from the surface to the safety of their nests under the ground. Burrowing animals serve a useful function: their activities help to loosen the soil and let in air and water.

Ponds are hollows in the ground that become filled with water. Ponds, like the soil, provide shelter for a wide range of plants and animals that spend all or part of their lives beneath the Earth's surface.

• Rabbits often live in underground burrows during cold winter months. They usually do not dig their own burrows but move into ones abandoned by other animals.

• Moles spend most of their lives underground. They use their front claws like shovels to dig through the soil in search of food. Moles can burrow through up to a hundred yards of soil in a day.

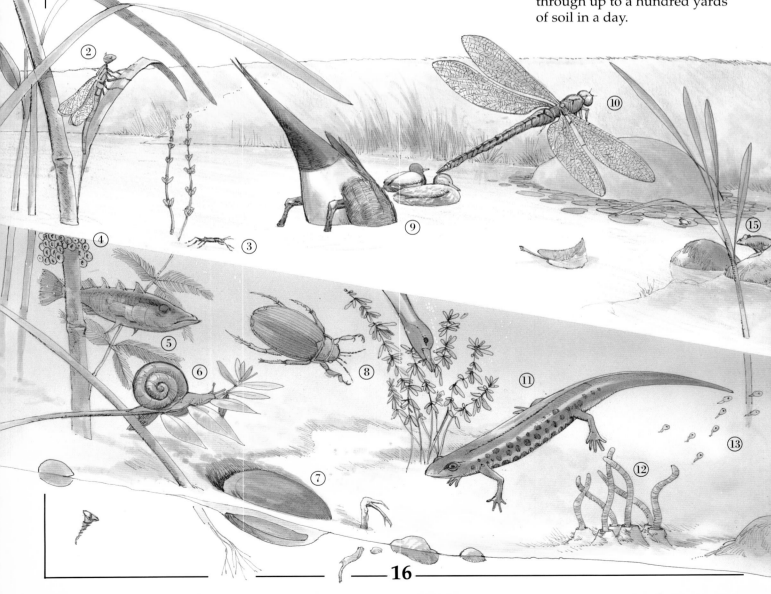

KEY

1. Mole
2. Caddis-fly
3. Pondskater
4. Frog eggs
5. Stickleback
6. Snail
7. Freshwater clam
8. Water beetle
9. Ducks
10. Dragonfly
11. Newt
12. Tubifex worms
13. Tadpoles
14. Frog
15. Water rat
16. Shrew
17. Molehill
18. Mole
19. Earthworm
20. Rabbit nesting chamber
21. Shrew
22. Mouse nest
23. Rabbits

# CABLES, DRAINS, AND SEWERS

① Although people rarely descend beyond the upper layer of the Earth's surface, the buildings we live and work in are usually serviced by a network of underground facilities. The ground beneath the streets of a city is interlaced with cables, drains, and sewers.

When we turn on a faucet, the water that flows out has been transported from local reservoirs through water mains that run underneath the city streets. Wastewater flows through drains into sewer systems that lead to sewage plants. The gas that fuels our stoves and furnaces flows through underground pipes. Telephone, electrical, and television cables are sometimes buried beneath the streets or suspended along underground pipes or tunnels.

• Rats living in underground sewers and tunnels can be a nuisance. They sometimes bite through cables and even lead pipes.

②

**18**

• The ancient Romans built underground water supply systems and sewers. One of the main sewers of Rome, called Cloaca Maxima, is still in use today.

**19**

# CAVES

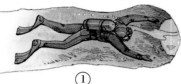

Caves are usually found in limestone areas. Limestone is made of a mineral called calcite. It is a hard rock that is easily dissolved by a weak acid, called carbonic acid. Carbonic acid is formed when rain dissolves carbon dioxide from the air and soil. Over thousands of years, rainwater can eat away at layers of limestone, creating a network of tunnels and caves. The rainwater trickles through cracks in the rock, enlarging the cracks and finding new paths between the layers of stone. The paths widen into tunnels. The tunnels crisscross and grow into rooms, or chambers.

KEY
1. Spelunker (explorer of caves and underground rivers)
2. River
3. Limestone cliffs
4. Sinkhole (hole in the limestone that leads to underground chamber)
5. Cave chamber
6. Underground river
7. Stalagmites
8. Stalactites
9. Sandstone
10. Sinkhole
11. Limestone
12. Spelunkers
13. River running into cave and joining underground river

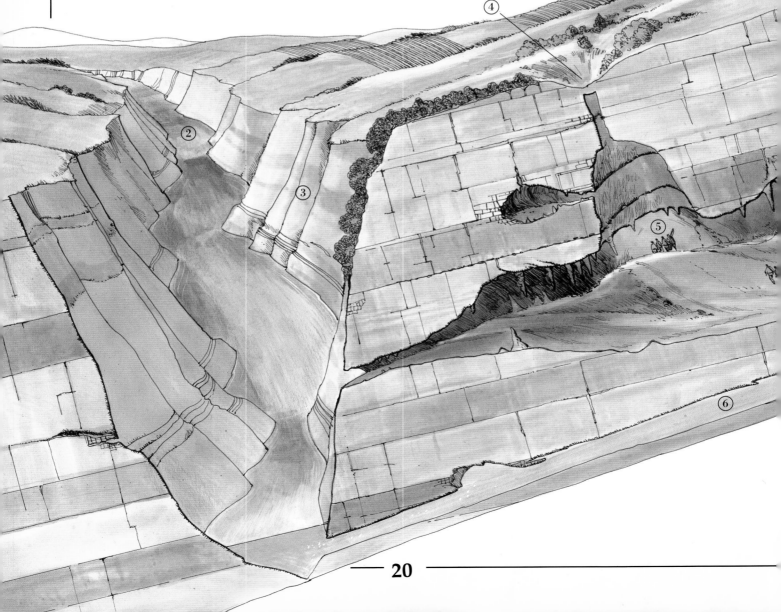

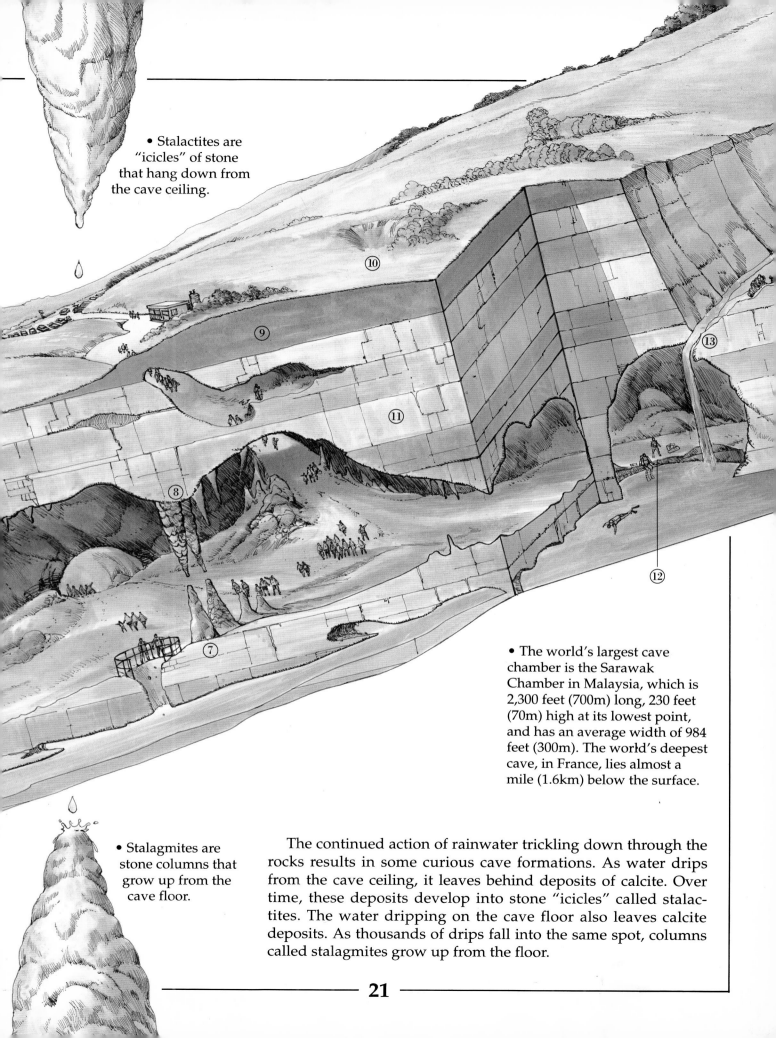

• Stalactites are "icicles" of stone that hang down from the cave ceiling.

• The world's largest cave chamber is the Sarawak Chamber in Malaysia, which is 2,300 feet (700m) long, 230 feet (70m) high at its lowest point, and has an average width of 984 feet (300m). The world's deepest cave, in France, lies almost a mile (1.6km) below the surface.

• Stalagmites are stone columns that grow up from the cave floor.

The continued action of rainwater trickling down through the rocks results in some curious cave formations. As water drips from the cave ceiling, it leaves behind deposits of calcite. Over time, these deposits develop into stone "icicles" called stalactites. The water dripping on the cave floor also leaves calcite deposits. As thousands of drips fall into the same spot, columns called stalagmites grow up from the floor.

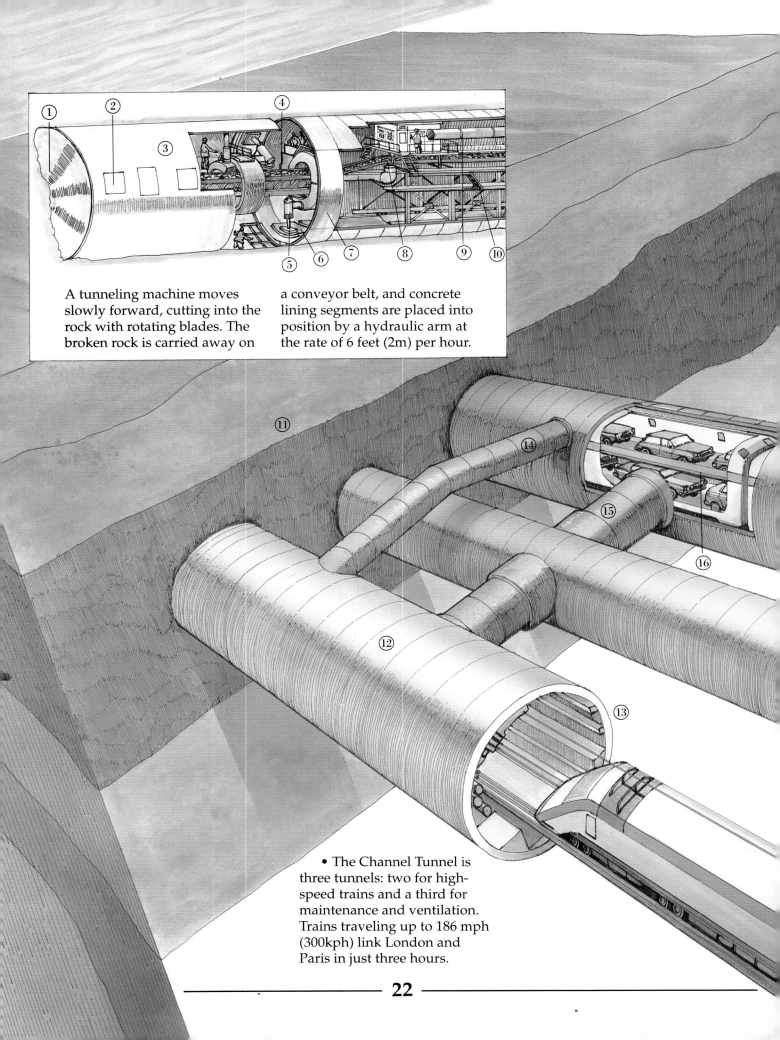

A tunneling machine moves slowly forward, cutting into the rock with rotating blades. The broken rock is carried away on a conveyor belt, and concrete lining segments are placed into position by a hydraulic arm at the rate of 6 feet (2m) per hour.

• The Channel Tunnel is three tunnels: two for high-speed trains and a third for maintenance and ventilation. Trains traveling up to 186 mph (300kph) link London and Paris in just three hours.

# TUNNELS

People first began to dig tunnels to mine minerals from the Earth and to carry water supplies. The tunnel systems of the ancient Greeks and later the Romans were marvels of engineering. Although power drills, digging machines, and explosives have made the task of building tunnels much easier, it is still a dangerous undertaking.

Many modern tunnels are built to carry traffic through mountains or beneath cities or rivers. The quickest way of tunneling is to dig a deep trench in the ground, line the sides with solid walls, build a roof, and then cover the whole structure over with soil and rock. This "cut-and-cover" method is used only when the tunnel is close to the surface. The other method is to dig a tunnel under the ground starting from vertical shafts at each end. This is more difficult because there is always the risk that the tunnel roof may collapse during construction.

• One of the most ambitious tunneling projects ever undertaken is the Channel Tunnel between England and France. It is 30.7 miles (49.4km) long, of which 23.6 miles (38km) are under the sea.

KEY

1. Cutting blades
2. Control cabin
3. Steel casing
4. Conveyor belt
5. Hydraulic arm
6. Lining the tunnel
7. Lining segment
8. Removing the broken rocks
9. Air supply
10. Walkway

11. English coast
12. Main tunnel for rail traffic only
13. Concrete outer casing
14. Relief ducts to reduce air pressure in front of moving trains
15. Cross passage, connecting main tunnels to service tunnel
16. Double deck railcar
17. Service tunnel for ventilation and maintenance

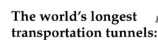

**The world's longest transportation tunnels:**

1) Seikan tunnel between the islands of Honshu and Hokkaido in Japan: 33.5 miles (54km)

2) Channel Tunnel connecting Britain and France: 30.7 miles (49.4km)

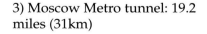

3) Moscow Metro tunnel: 19.2 miles (31km)

4) Simplon rail tunnel under the Swiss Alps between Italy and Switzerland: 12.4 miles (20km)

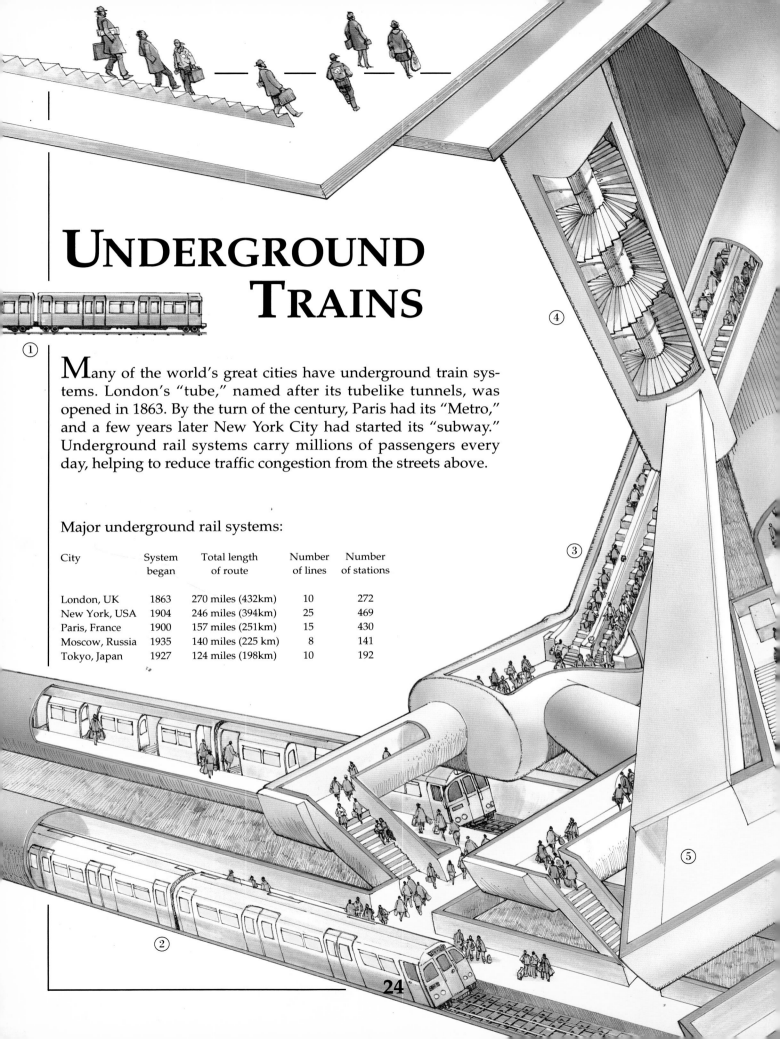

# UNDERGROUND TRAINS

① 

Many of the world's great cities have underground train systems. London's "tube," named after its tubelike tunnels, was opened in 1863. By the turn of the century, Paris had its "Metro," and a few years later New York City had started its "subway." Underground rail systems carry millions of passengers every day, helping to reduce traffic congestion from the streets above.

Major underground rail systems:

| City | System began | Total length of route | Number of lines | Number of stations |
|------|--------------|----------------------|-----------------|---------------------|
| London, UK | 1863 | 270 miles (432km) | 10 | 272 |
| New York, USA | 1904 | 246 miles (394km) | 25 | 469 |
| Paris, France | 1900 | 157 miles (251km) | 15 | 430 |
| Moscow, Russia | 1935 | 140 miles (225 km) | 8 | 141 |
| Tokyo, Japan | 1927 | 124 miles (198km) | 10 | 192 |

24

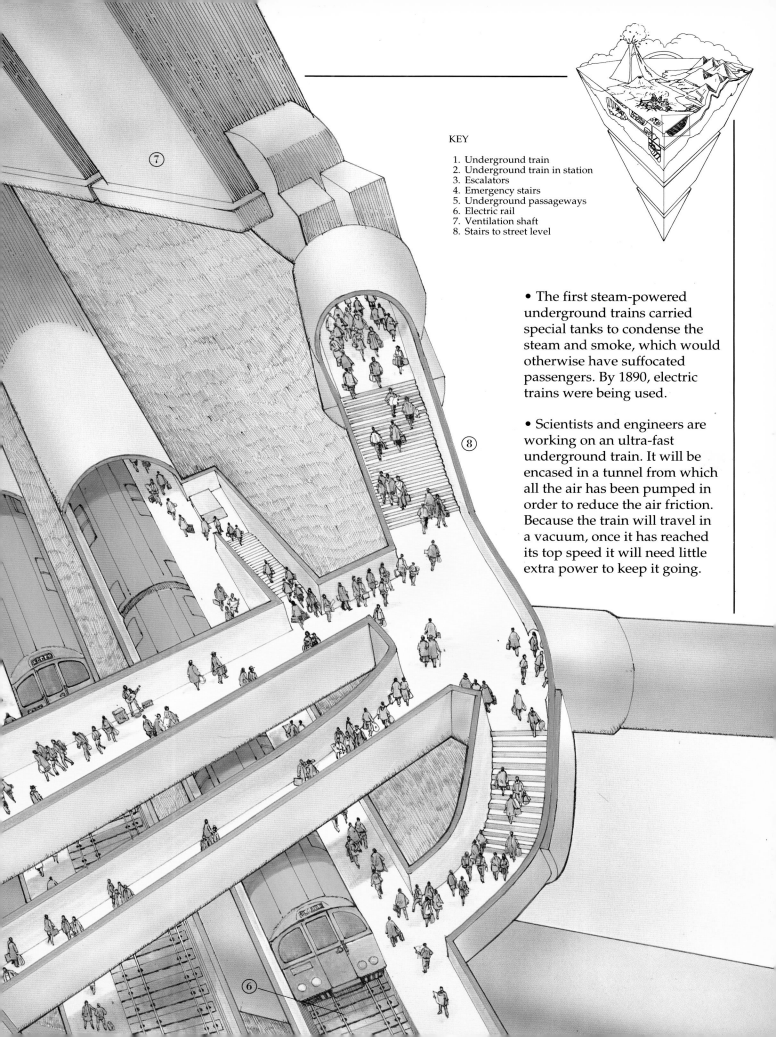

KEY

1. Underground train
2. Underground train in station
3. Escalators
4. Emergency stairs
5. Underground passageways
6. Electric rail
7. Ventilation shaft
8. Stairs to street level

• The first steam-powered underground trains carried special tanks to condense the steam and smoke, which would otherwise have suffocated passengers. By 1890, electric trains were being used.

• Scientists and engineers are working on an ultra-fast underground train. It will be encased in a tunnel from which all the air has been pumped in order to reduce the air friction. Because the train will travel in a vacuum, once it has reached its top speed it will need little extra power to keep it going.

# MINES

①

The Earth's rocky crust is made up of minerals. Metals such as copper, gold, and tin, which have been mined since ancient times, are minerals. One of the most valuable minerals stored in the Earth is coal. Although coal originated from organic material, it is usually called a mineral. Coal was formed from the remains of trees and other plants that lived about 300 million years ago when much of the land was wet and marshy. As the plants died, they fell into swamps and were partly decayed, forming a black substance known as peat. In time, deposits of mineral matter covered the peat beds, and the piled up weight compressed the moisture out of the peat and left behind hard coal.

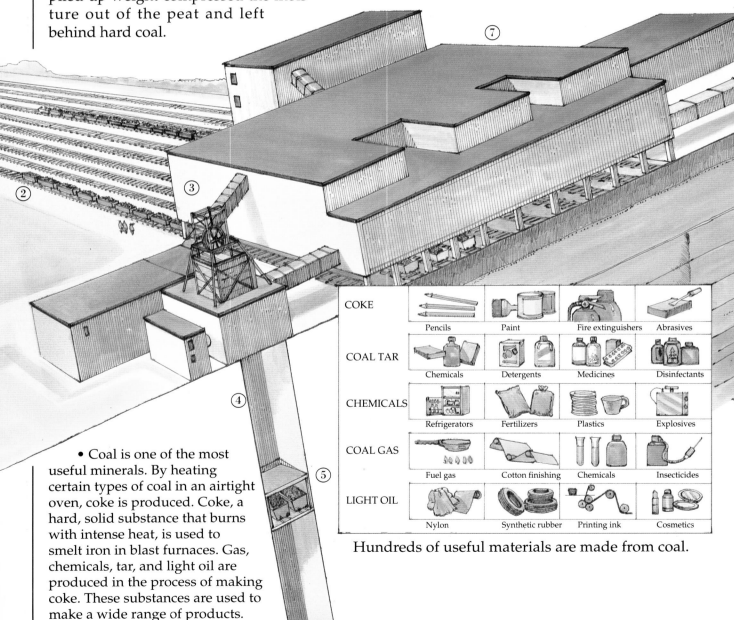

| COKE | | | | |
|---|---|---|---|---|
| | Pencils | Paint | Fire extinguishers | Abrasives |
| COAL TAR | | | | |
| | Chemicals | Detergents | Medicines | Disinfectants |
| CHEMICALS | | | | |
| | Refrigerators | Fertilizers | Plastics | Explosives |
| COAL GAS | | | | |
| | Fuel gas | Cotton finishing | Chemicals | Insecticides |
| LIGHT OIL | | | | |
| | Nylon | Synthetic rubber | Printing ink | Cosmetics |

• Coal is one of the most useful minerals. By heating certain types of coal in an airtight oven, coke is produced. Coke, a hard, solid substance that burns with intense heat, is used to smelt iron in blast furnaces. Gas, chemicals, tar, and light oil are produced in the process of making coke. These substances are used to make a wide range of products.

Hundreds of useful materials are made from coal.

Coal has been mined for centuries. The first coal mines in Europe began in the thirteenth century. Vast reserves of coal still remain underground, but many of the easily accessible deposits have been mined. Today, miners use high-powered machinery that chews into the coal face and scoops up the pieces with a loading claw. Then it deposits them into waiting shuttle cars or onto conveyor belts that carry them to the surface.

• The deepest point that humans have reached below the surface is the bottom of the Western Deep Levels Mine, a gold mine near Johannesburg, South Africa. It is about 12,000 feet (3,800m) deep. Giant refrigeration units are needed to keep the air temperatures cool enough to be tolerable for the miners. The temperature of the rocks at this depth is over 120°F (50°C), hot enough to burn bare hands.

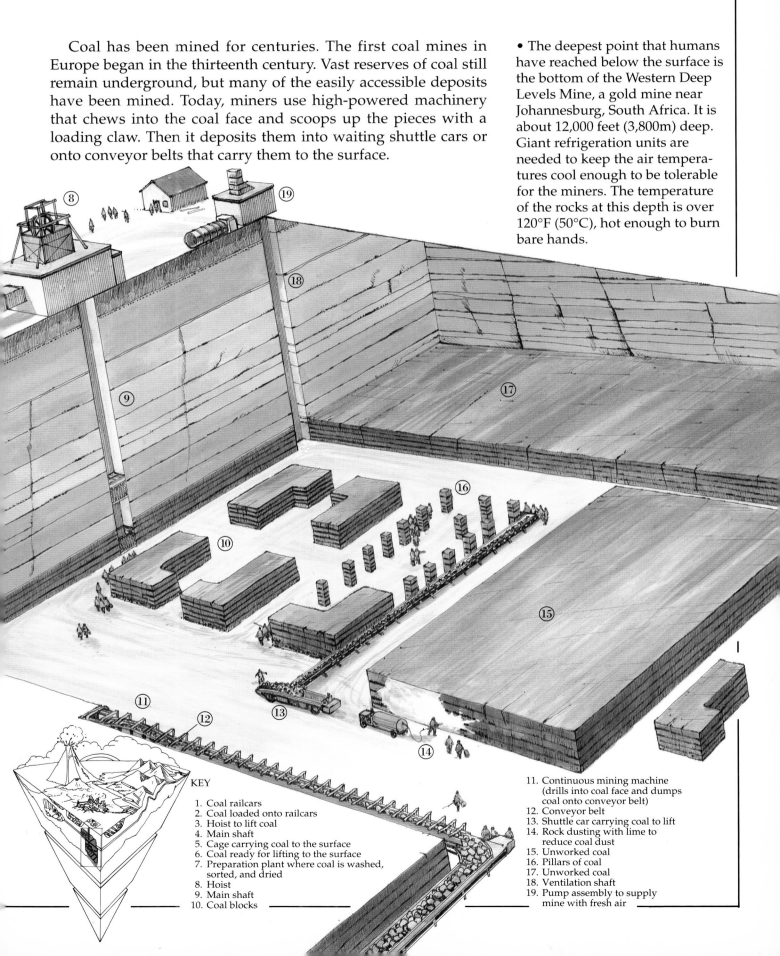

KEY
1. Coal railcars
2. Coal loaded onto railcars
3. Hoist to lift coal
4. Main shaft
5. Cage carrying coal to the surface
6. Coal ready for lifting to the surface
7. Preparation plant where coal is washed, sorted, and dried
8. Hoist
9. Main shaft
10. Coal blocks
11. Continuous mining machine (drills into coal face and dumps coal onto conveyor belt)
12. Conveyor belt
13. Shuttle car carrying coal to lift
14. Rock dusting with lime to reduce coal dust
15. Unworked coal
16. Pillars of coal
17. Unworked coal
18. Ventilation shaft
19. Pump assembly to supply mine with fresh air

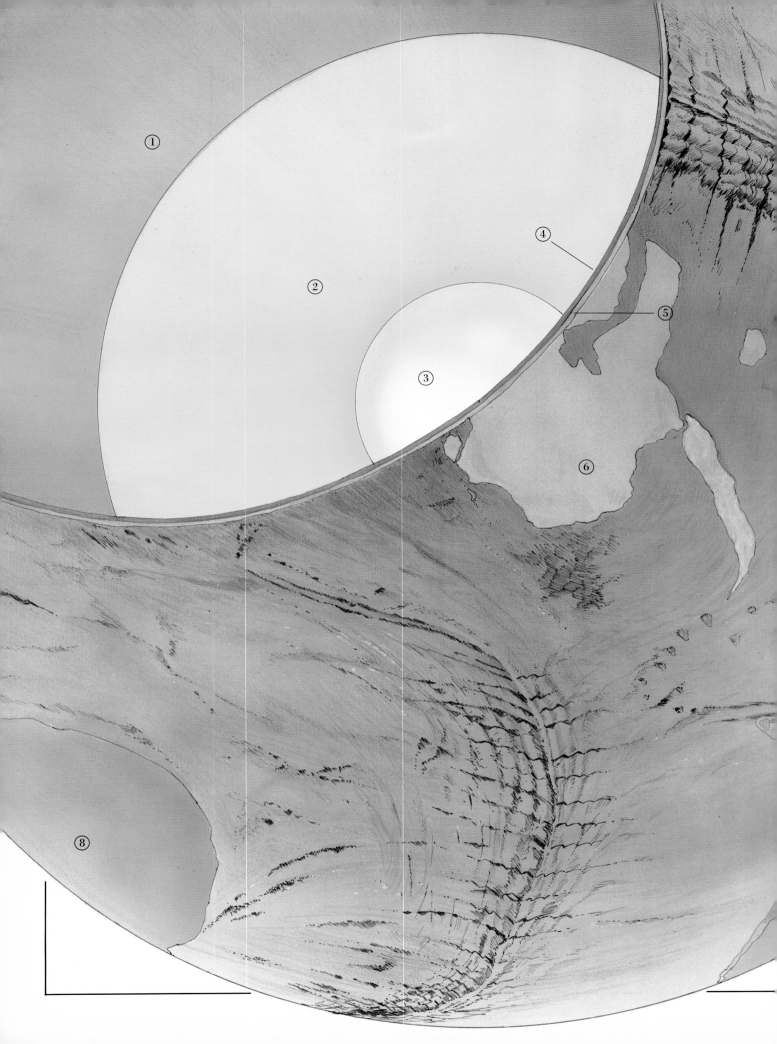

# TOWARD THE CORE

All life on Earth exists within the top layers of the crust. Even the deepest mines have barely penetrated it. Although we are unable to visit the deep interior of the Earth or even place measuring instruments there, scientists have been able to learn about its interior structure by studying the path of seismic waves as they travel through the planet. At certain points the waves suddenly change direction and speed in response to changes in the composition of the material they are passing through. These points mark the boundaries between the different layers of the Earth — the crust, mantle, and core.

The thickness of the Earth's crust varies from only 5 miles (8km) under the oceans to an average of 25 miles (40km) under the continents. Beneath the Earth's crust is the mantle, which extends (from the base of the crust) to a depth of about 1,800 miles (2,900km). The Earth's core is divided into two parts: the liquid outer core, which is about 1,400 miles (2,250km) thick, and the solid inner core, which is about 850 miles (1,370km) thick.

The temperature at 250 miles (400km) below the surface, in the Earth's mantle, is estimated to be over 1,600°F (870°C). At the core it is even hotter, perhaps as much as 9,000°F (5,000°C).

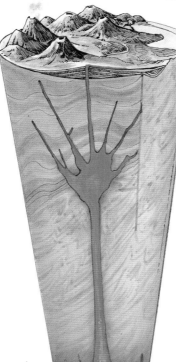

KEY

1. Mantle
2. Outer core
3. Inner core
4. Boundary between mantle and crust
5. Crust
6. Antarctica
7. South America
8. Africa

⑦

• Part of the mantle is semi-molten. Magma rises up and collects in a magma chamber in the upper layer of the mantle.

• Because liquid magma is lighter than solid rock, it rises, forcing its way to the surface through weak spots in the upper mantle and crust.

# GLOSSARY

**Bacteria**
Microscopic, single-celled organisms found in soil, water, air, and the human body. Many types of bacteria bring about chemical changes, such as the decay of organic matter and the building up of nitrogen compounds in the soil.

**Carbon dioxide**
A colorless, odorless gas that is somewhat heavier than air.

**Carbonic acid**
A weak, colorless acid that forms when carbon dioxide is dissolved in water.

**Continental plate**
One of the plates of the Earth's crust that supports a landmass or continent. The Earth's land surface is made up of seven continents. These are Asia, Africa, North America, South America, Europe, Australia, and Antarctica.

**Core**
The core of the earth is made up of two distinct parts: a liquid outer core and a solid inner core. The core extends from the base of the mantle to the Earth's center, a distance of 2,200 miles (3,550km), and accounts for about 17 percent of the Earth's volume.

**Crust**
The thin outer layer of the Earth. It makes up only about 0.6 percent of the planet's volume. The thickness of the crust varies from just 5 miles (8km) under the oceans to as much as 50 miles (80km) under mountain ranges.

**Equator**
An imaginary line around the Earth halfway between the poles.

**Erosion**
The gradual wearing down of materials the Earth is made of. Wind, rainfall, oceans, running water, or ice can cause erosion.

**Fault**
A crack in the Earth's crust.

**Fissure**
A long crack, or fault, through which gas and magma find their way to the Earth's surface.

**Insect**
A small invertebrate animal with six legs, two or four wings, and a body divided into three segments.

**Lava**
Molten rock that flows on the Earth's surface.

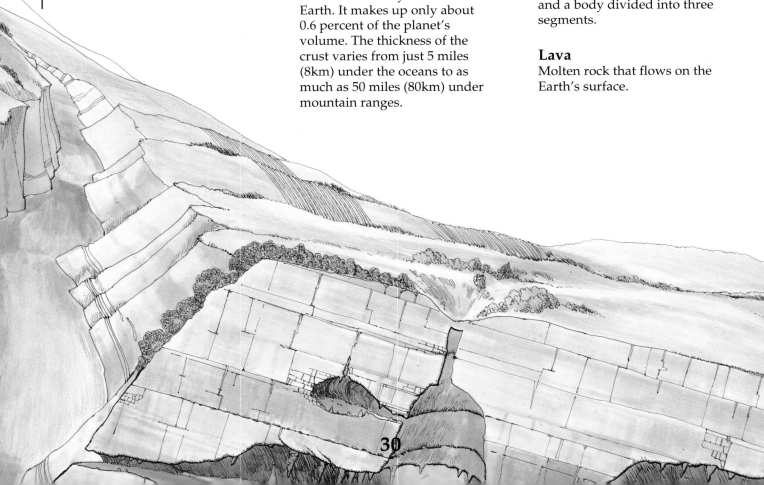

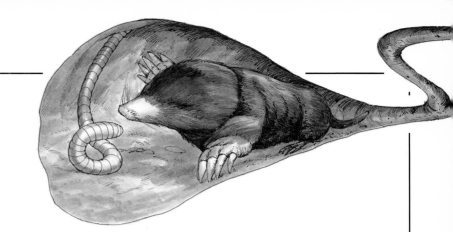

## Limestone

A rock that was formed millions of years ago. Some limestones were formed when tiny sea animals died and fell to the bottom of the sea and formed a layer of limy ooze. Over time, as the ooze was covered with other deposits, it compressed and formed limestone. Movement of the Earth's plates has brought deposits of limestone to the surface. Limestone can be dissolved by weak acids.

## Magma

Molten rock inside the Earth. When it reaches the surface it is called lava.

## Mammal

A warm-blooded animal that is usually covered with hair or fur and feeds its young on milk.

## Mantle

The Earth's mantle extends from the base of the crust to a depth of about 1,800 miles (2,900km). The mantle accounts for about 82 percent of the Earth's volume. It is mostly solid but contains a semi-molten layer.

## Mineral

An inorganic substance occurring naturally in the Earth. Each mineral has a specific chemical composition and distinctive physical characteristics. The term mineral is sometimes applied to a substance in the Earth that is of organic origin, such as coal.

## Molten

Melted.

## Oceanic plate

A plate of the Earth's crust that lies underneath an ocean.

## Organic

Derived from living organisms.

## Planet

One of the bodies that revolves around the sun. The Earth is one of the planets.

## Plate

One of the several large rocky segments that make up the surface of the Earth.

## Reservoir

A large natural or artificial lake in which water is collected and stored to supply a given area.

## Seismic waves

Tremors or vibrations caused by an earthquake.

## Sewer

A tunnel that carries dirty water and human wastes.

## Tunnel

An underground passage.

## Volcano

An opening in the Earth's surface through which magma, gases, and rock fragments are emitted. Mountains that form around such openings are also called volcanoes.

# INDEX

References in bold type refer to illustrations.

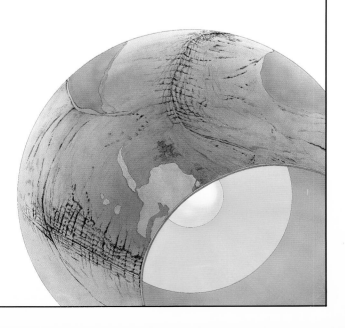